Health, Peace and CONFIDENCE

Thriving in the Midst of Adversity

LAURA R. FERNÁNDEZ, MPH, BSN, RN

Leader in the Valuable Woman Community

Editorial Güipil

For more resources, visit:
EditorialGuipil.com

Health, Peace & Confidence
Thriving in the Midst of Adversity

Published by Editorial Güipil
Miami, FL - Winston-Salem, NC. United States of America

Editorial Güipil

Editorial Güipil
First Edition 2025
EditorialGuipil.com

ISBN: 978-1-953689-96-2
Category: Wellness / Christian Life / Inspiration

«Everything the world offers cannot heal the broken heart, nor give peace to the spirit, nor dispel unease, nor banish disease. Fame, genius, and talent are powerless to cheer the sorrowful heart or restore a wasted life.
The life of God in the soul is man's only hope.»

—Ellen G. White.

Acknowledgment

My deepest gratitude to my husband Freddy, because without his constant support and immense patience during the long nights when ideas would mix and tangle together, this book would not have been completed.

I am also grateful that God allowed me to work for more than 36 years in the healthcare system, where I was able to closely see, listen, understand, empathize, and most of all, assist patients in the midst of need, pain, despair, discouragement, sadness, depression, anxiety, and even imminent death. This, along with my Christian principles, led me to search beyond conventional medicine and social trends to try and reveal a safer and more effective path that would lead the reader to enjoy a full and healthy life, a life of blessing and success.

Dedication

To every person who, with sincerity and willpower, decides to change poor nutrition, an intemperate lifestyle, and all bad habits in order to restore their physical and mental health, and thus regain their spiritual growth and connection with God.

Content

Introduction

Come closer. Let go of your worries for a moment. Take a deep breath in, and exhale slowly while counting to eight. That's it. Very good. Now that I have your attention, I'd like to ask you something: How much would you be willing to pay, do, or invest to enjoy good health, not just physical, but also mental and spiritual? Don't worry, this is a rhetorical question. Nowadays, bookstores, social media, and advertising channels are full of offers and new suggestions promising health under the guise of a new pill, a new exercise routine, a new diet, or a new kind of esoteric technique that might work for a few weeks or months. Unfortunately, most people end up right back where they started, or even worse off than before.

It's possible that you, a loved one, or someone you know is right now trying to survive amidst a sea of failed attempts, a pile of impossible-to-follow instructions, or perhaps a mountain of religious or spiritual advice that simply hasn't worked. You may still feel weak from illness, weighed down by depression or anxiety, or trapped by past traumas and destructive feelings of guilt.

Don't worry, I'm not going to sell you a miracle product. Nor will I recommend a specific workout plan requiring expensive equipment or endless hours at the gym. I won't offer you mantras, repetitive prayers, or elaborate spiritual disciplines focused on an egocentric inner force that can be, frankly, useless or even harmful. But there is something I want to share with you. And that is the possibility and the inspiration to discover a lifestyle that can change your current situation. A lifestyle that can restore the long sought for harmony between physical, mental, and spiritual health.

Most people in the 21st century are living halfway. They are disconnected within themselves, split between these three dimensions, and that prevents them from living fully. This disconnection keeps them from enjoying life's beauty and limits their ability to overcome adversity. They face challenges like illness, injustice, sadness, and pain largely alone. Through this book, I want you to discover that there is hope to triumph in life. There is comfort for those who suffer, who grieve, and who weep. Today is your chance to draw near to the source of love, health, and the power to overcome. Tomorrow is another day, bringing fresh opportunities to adjust your course and make the right choices.

You have the power to choose among the options before you, but with that power comes responsibility. It's time to stop blaming others for your situation or hiding behind painful events from your past. I am in no way trying to minimize your personal experiences. Rather, I want to help you break free from the vicious cycle you've fallen into, or might fall into, if you keep dwelling on the negative. That mindset will only drain the energy and resources you need to move forward and succeed in life.

I tell you this from personal experience. As someone who came from a large, dysfunctional family, though rooted in Christianity, and after over 36 years of working in the healthcare system, I can assure you that the solution to regaining your health, peace of mind, and the joy of living in tune with your spiritual side is simple, yet effective. It's far more practical than you imagine and carries eternal rewards.

DON'T MISS THIS OPPORTUNITY!

I invite you to explore each of these three dimensions with me. Understand their qualities, the care they demand, and the precautions needed. By doing so, you'll be empowered to make the necessary changes in your life and embrace a fullness marked by meaning and purpose, one where your mind and body finally rest in the loving arms of a Savior who gave His life so that we might be restored to the original place for which we were created in the universe.

The Bible says in 1 Thessalonians 5:23–24:

"May God himself, the God of peace, sanctify you through and through. May your whole spirit, soul and body be kept blameless at the coming of our Lord Jesus Christ. The one who calls you is faithful, and he will do it."

And also in Isaiah 40:31:

"But those who hope in the Lord will renew their strength.

They will soar on wings like eagles;
they will run and not grow weary,
they will walk and not be faint."

With all my heart, I hope this becomes a reality in your life.

Come with me through the following chapters and discover how beautiful it is to live in harmony with God, with your inner self, and with others.

CHAPTER 1:
CREATED IN HARMONY

THE CLOSE RELATIONSHIP BETWEEN NATURE AND THE HUMAN BEING

The earth, humanity, and the world we live in are filled with beauty and extraordinary wonders. It is truly uplifting to gaze at the blue sky, a starry night, the song of birds, and the vastness of the sea. The earth, humanity, and the world we inhabit are filled with beauty and extraordinary wonders. It is truly uplifting to gaze at the blue sky, a starry night, the song of birds, and the vastness of the sea.

I grew up on a beautiful Caribbean Island and still feel a deep connection to the sea, the beach, and the lush greenery of the tropics. I find that listening to the steady rhythm of the waves is one of the best ways to soothe a mind troubled by the pressures of daily life and a body worn by routine. The same can be said for the gentle flow of a stream winding between rocks or the refreshing patter of rain on a warm summer afternoon.

As a professional nurse with many years of experience and graduate studies in public health, I have witnessed the progress medicine has made in holistic health. Recent research confirms the value of combining conventional medical treatment with natural remedies.

Don't be surprised if, on your next visit to your primary care doctor, a conversation about managing stress and anxiety leads to a prescription for spending two weeks in the countryside or forest. Studies show that the brain can regenerate its electrical activity and improve neuronal connections simply through the contemplation of, and proximity to, trees, reducing anxiety and fostering a sense of well-being.

Nature, even after enduring so many years under the harsh hand of humanity, still preserves much of its indescribable beauty. Within it, we can hear the voice of God and see reflected what is often called God's second book, after His Holy Word. Romans 1:20 tells us:

> "For since the creation of the world God's invisible qualities, His eternal power and divine nature, have been clearly seen, being understood from what has been made, so that people are without excuse."

Through nature, we can lift our thoughts to a loving and merciful Creator who delights in showering countless blessings upon His children. The biblical narrative presents humanity as the crowning achievement of divine creation within a magnificent original plan. We were given an incredible body, capable of complex biological functions, and a powerful mind that analyzes, reasons, loves, and feels. Above all, we were created in His image, with a spiritual nature that connects us directly to our Maker. We will explore this further in the chapters ahead.

Truly, nature offers a spectacular daily display of diversity that can satisfy the needs of any willing heart. Trees are a living example of this diversity. When was the last time you paused to observe the beauty of a tree or a forest? As you looked closely, did you notice its unique qualities and attributes? Did you examine features such as the type of wood, whether (ornamental, fruit-bearing, or flowering) the shape of its leaves, or its growth cycle? These, and countless other questions, could be asked about any tree. Imagine the richness of study material available for botany and dendrology, the science of trees and shrubs. What about those dense enough to provide shade for a weary traveler?

One place on my travel wish list is the famous Redwood National and State Parks in northern California, USA. I've seen them in photos, movies, documentaries, and more, but I would love to visit them someday. There stand the majestic sequoias, the tallest trees on earth. The tallest among them, discovered in 2006, is called Hyperion. In 2019, it was measured at an impressive 116.07 meters (about 380 feet) tall. Its wood is a rich reddish color. It would be a delight to see the contrast between this towering giant and my own petite height of 5'3".

Of course, scientists go far beyond the simple observations of an ordinary person. Professionals who dedicate their time to studying these trees take into account even the smallest details to maintain meticulous records of each species and type. As you may have noticed, trees are an object of my admiration. I find them captivating, some are imposing, others display stunning colors, some have amusingly shaped leaves, and others still boast enormous roots and a delightful aroma. One detail that catches my attention is that in the Bible, the righteous are compared to trees in a beautiful divine promise found in Psalm 1:3:

"That person is like a tree planted by streams of water, which yields its fruit in season and whose leaf does not wither, whatever they do prospers."

We cannot deny that nature is an open book for anyone eager to learn about beauty. Now, I invite you to shift your paradigm and imagine the vast field of study available to both professionals and amateurs when it comes to the human body, the crowning achievement of divine creation, and the astounding capacities, inner workings, and secrets held by the brain and the other organs of this marvelous creation. It was precisely this curiosity about these processes that led me to study nursing. I had a deep need to know, explore, and experience firsthand not only the extraordinary nature of the human body but also the magnificence, perfection, power, and supreme love of the One we call God, our Creator and Sustainer.

Life is a wonderful and precious gift that few truly understand. Every birth, no matter how commonplace it may seem, is a sublime miracle, encompassing months of countless processes, divisions, specializations, complex functions, and incalculable precision, all of which can only occur under the wise and essential direction of a loving and divine Master. In truth, we were created in harmony with everything good and perfect that comes from God, but this harmony was disrupted when the virus of sin entered and wreaked havoc on God's original plan. As a result, human life has deteriorated in all three dimensions: physical, mental, and spiritual. What was created to live forever, with the passing of time, and amidst both failures and victories, the best among us now reach an average lifespan of around 80 years in developed countries, with a few individuals surpassing 100. Years ago, a study was published about the so-called Blue Zones, regions where people commonly live beyond 100 years.

It is striking that in the book of Genesis, after the story of the global flood, God established that human life would be limited to around 120 years. Still, even in the secular realm, doctors and scientists constantly work to combat the multitude of diseases and conditions that affect human health and vitality during that time before the inevitable end.

The various branches of medicine labor persistently to fight against anything that disrupts bodily homeostasis, or, in simpler terms, the harmony we enjoy when we are healthy. Major medical research organizations, hospitals, urgent care centers, and primary care offices work tirelessly to alleviate physical pain, heal or eradicate disease, and stabilize any trauma affecting our physical or mental well-being. However, although they know and work diligently to restore homeostasis in these two areas, they understand there is a third, spiritual, which they do not enter, leaving it to the church or to the individual. Thus, the church steps in to address spiritual afflictions. It is often considered a vast hospital where each person can approach God and find healing for a tired and troubled spirit. But it's not just the church, this entire world is a vast and complex hospital where human beings constantly face the battle between life and death, between health and disease. Regardless of whether or not you believe in God or understand or accept His involvement in the affairs of mankind, life itself reveals that this is true.

MY EXPERIENCE

I would like to share with you a personal experience. Let's go back to a quiet morning in January of 2007. Along with my husband and in-laws, I arrived at a local hospital where I was scheduled for a routine procedure: the removal of my gallbladder. As a nurse, I knew this procedure is commonly performed at outpatient surgery centers, as it is not considered high risk. These days, you're operated on in the morning and can go home that same afternoon. However, the surgeon had a scheduling conflict at the outpatient center and had to book the procedure at the hospital where he also practiced.

I received instructions from the medical team and said goodbye to my husband and family before the first sedatives were administered.

Of course, as believers, we offered a prayer to God, asking for His care and protection during the procedure. Once in the operating room, I was placed under general anesthesia and lost awareness of everything happening around me. The doctor turned to my husband and said:

—See you in 40 to 45 minutes.

But only about 20 minutes had passed when the doctor came out, looking very serious, and motioned for my husband to come over. Once he approached, the doctor asked if he was alone. My husband replied that he was with his parents. The three of them were taken to a small chapel, where the doctor informed them that everything was okay, but he had had to stop the procedure because I had experienced a reaction to one of the anesthesia medications. He told them the medical team was working on me, and that I might be seen in an hour or two. My husband later told me that his world crumbled the moment the doctor led them to that chapel, but he felt some reassurance when he heard the words, "everything is under control."

Now I'd like to share the details of what actually happened in the operating room before the doctor spoke to my husband. The diagnosis and reason the procedure had to be halted was not just a mild reaction to the anesthesia, I had experienced an anaphylactic shock.

Anaphylactic shock is a severe, systemic reaction to a substance to which the person is allergic. Complications can include brain damage or kidney failure. Without immediate treatment, death can occur within fifteen minutes. This condition occurs in fewer than 1% of cases, in other words, it's clinically rare, but carries a high mortality rate.

By the time my husband was allowed to see me, I was still under the effects of multiple medications. My face was red and my entire body extremely swollen. He later told me how frightened he was to see me that way, my eyes were nearly swollen shut.

As part of my recovery, I had to undergo a cardiac catheterization the next day. The doctors wanted to confirm that no damage had been done to my heart from the lack of oxygen during the efforts to regulate my heartbeat and resuscitate me. They also ran tests to ensure there was no kidney damage or lasting neurological effect.

In short, both from personal experience and professional knowledge, I can testify that the fact I'm writing these words sixteen years after that incident is proof that I was the recipient of a miracle.

During the emergency, I was unconscious. I remember nothing of what happened. I simply opened my eyes and saw myself surrounded by medical staff who were watching me closely and saying:

—Hello! We're so glad you're okay!

As the hours passed, I began to learn more about what had happened.

Once I returned home and resumed my responsibilities at home, at work, and in church, all I could do was thank God for this new chance at life. It's important to note that there was no way for the doctors to know about the severity of my allergy, I didn't even know it myself… but God did!

GOD'S CONSTANT AND FAITHFUL CARE

Some may think it was all just a coincidence, that whatever is meant to happen simply happens, and that's the end of it. If you're one of those people, I invite you to pause and reflect on a few key and meaningful points from my experience, or from the life of anyone who has faced similar circumstances, where relationships and human connections suddenly take on deeper significance.

It was no coincidence that the specific group of health professionals happened to be present during my emergency. It was no coincidence that the surgeon couldn't perform the procedure at the outpatient center and instead had to schedule it at the hospital, where the full spectrum of medical resources and specialties were available. And it certainly wasn't a coincidence that the surgery hadn't yet begun, allowing the doctor to respond immediately to the unfolding crisis. Jesus was present in that operating room.

His holy angels guided the hands of those medical professionals so that they acted with precision, skill, and wisdom. Because of that divine intervention, I was able to leave the operating room unharmed. Long before I was placed under anesthesia, a Bible verse kept echoing in my heart, Isaiah 41:10:

> "Do not fear, for I am with you;
> do not be dismayed, for I am your God.
> I will strengthen you; I will help you;
> I will uphold you with My righteous right hand."

My part was to trust in God and call upon Him for protection in prayer. My Heavenly Father took care of the rest. It didn't matter whether the doctors or nurses were believers, what mattered was that a daughter of God was in need, and God did not delay in responding to her cry.

Some might ask whether God only helps those who belong to Him. But who are God's children? The answer is simple: we are all God's children by creation, but only those who accept His gift of grace through Jesus Christ become His children by redemption. Romans 6:23 tells us:

"For the wages of sin is death, but the gift of God is eternal life in Christ Jesus our Lord."

Let me be clear: God is just, merciful, powerful, and holy, and He delights in pouring out His perfect love upon His creation. As John 3:16–17 assures us:

"For God so loved the world that He gave His one and only Son, that whoever believes in Him shall not perish but have eternal life. For God did not send His Son into the world to condemn the world, but to save the world through Him."

God's Plan for Restoration

God created us in perfection. Yes, His original plan was interrupted by the entrance of sin, but He already had the solution in place. Though the virus of sin brought desolation, disease, and death, through His promise, and in Christ Jesus, we can find peace, joy, healing, and life. We can experience the original harmony that our first parents, Adam and Eve, once enjoyed.

Let me offer an illustration: the world is like a vast hospital, and each of us is battling the effects of illness and its consequences, whether physical, mental, or spiritual. My time working in an emergency room helped me see the human condition in this light. In the ER, anyone may walk in, whoever wants to, needs to, or is able.

In the United States, two well-known laws, —COBRA and EMTALA—, regulate emergency services. These laws require that anyone presenting with a serious condition must receive care, or at least be stabilized, regardless of legal status, race, age, gender, or ability to pay.

But here's a crucial point: the patient, or someone responsible for them, must accept the offered care. In cases where the patient is unconscious and no legal guardian is present; the medical team is obligated to preserve life to the best of their ability.

One sobering reality of emergency room work is this: we are all equal in the face of tragedy, illness, poor choices, and especially, death.

Spiritually, it is the same. No matter our condition, whether we believe in God or not, God is still God. He continues to love us, to offer restoration, and to invite us back into the inner harmony we were created to enjoy. The Bible expresses this beautifully:

Matthew 5:45 says:
"That you may be children of your Father in heaven. He causes His sun to rise on the evil and the good, and sends rain on the righteous and the unrighteous."

And in Numbers 23:19:
"God is not a man, that He should lie,
nor a son of man, that He should change His mind.
Does He speak and then not act?
Does He promise and not fulfill?"

In other words, God always draws near, to love, to care, and to bless. The evil, pain, despair, sickness, and injustice we encounter, these do not come from God. They originate from the enemy of God, who from the beginning has sought to remove Him from our lives, distort our understanding of His character, and cut us off from the true source of life and health.

Because Satan cannot rival God directly, his strategy is to wound, harass, and deceive humanity. He brings misery and suffering and then whispers that God has abandoned us or doesn't care.

And here is where many fall into error, believing that tragedy, illness, and hardship are punishments from a harsh and angry God. But this view stands in direct opposition to the tender, loving character of our Creator.

Consider this: the World Health Organization defines health not simply as the absence of disease, but as "a state of complete physical, mental, and social well-being." This definition points us back to the original harmony we were meant to enjoy, between body, mind, and spirit.

Human beings were created for four divine purposes:

- To be loved
- To love
- To serve
- And to reign

I invite you to explore these purposes through the lens of a fulfilled life, a life radiant with physical health, anchored in mental peace, and elevated by spiritual transformation. This is a life lived in total harmony, with God once more at the center as our Creator, Sustainer, and Savior.

Jesus declared in John 10:10:
"The thief comes only to steal and kill and destroy; I have come that they may have life, and have it to the full."

Here is the good news: if we surrender ourselves into the hands of our Maker, He as the Divine Potter, can reshape us into something new and whole. God has marked out a safe and certain path so that His children might walk toward total restoration.

Trust me, when you choose that path, the results will surprise you.

CHAPTER 2
A MASTERPIECE

THE WONDERFUL HUMAN BODY

One of the most beautiful truths found in the Bible is the declaration that human beings were created in the image and likeness of God (Genesis 1:26). As the crowning act of creation, God brought into existence two majestic beings, male and female, whom He called Adam, or humankind (Genesis 5:2). These beings were endowed with qualities and characteristics far superior to those of any other created organism. They bore the seal of heavenly glory and radiated divine royalty.

Our present condition makes it difficult to imagine the potential of these perfect beings in their original form. Yet we can catch glimpses of that distant majesty simply by observing the greatness and capabilities of the human body today, even after thousands of years of wear and damage brought on by sin.

When was the last time you paused in front of a mirror, really observed yourself, and reflected on your abilities and design? Did you notice the detail in your eyes, the symmetry of your hands, the expressiveness of your face, the uniqueness of your teeth, the movement of your fingers, or the distinctiveness of your feet? Every day, you perform countless tasks or motions without even thinking, because your body does them automatically.

Imagine, for example, that you're lying in bed and decide to get something from the kitchen. Without much thought, you get up, place your feet on the floor, turn toward the edge of the bed, pass the dresser or nightstand, walk through a partly open door, cross the living room, and arrive in the kitchen. All of this happens seamlessly, with your brain guiding the journey in milliseconds by retrieving information from memory because it's a route you've taken many times before. Your body calculates the energy required for each movement. Your muscles tense or relax accordingly. Your eyes, via the optic nerve, transmit visual data to your brain, which then measures the distance between you and the surrounding objects. Your nervous system sends the necessary electrical impulses to coordinate the entire process. All this occurs without you having to consciously decide each step.

Did you know your heart and lungs are part of the autonomic nervous system? That means you cannot consciously control them to the point of stopping them at will. These systems operate independently. Your heart began beating in the womb; your lungs activated at birth when they expanded for the first time due to the pressure change between the amniotic sac and the outside world.

Consider also your extraordinary filters, your kidneys and liver. Or your integumentary system, your skin, which not only gives you form and beauty, but also serves as your first line of defense against pathogens. The human body is a marvel of precision and complexity.

If you want to dive deeper, you might explore the digestive system with its intricate connections to the endocrine, circulatory, and lymphatic systems. Or consider the immune system, our body's alert mechanism and unmatched defense army. It is a system of astonishing precision, speed, and adaptability. Its ability to identify and respond specifically to countless external and internal threats is nothing short of miraculous. When we get sick, it's because the body has already exhausted every possible effort to contain or eliminate the invading threat.

Understanding the complexity of these processes helps us appreciate just how uniquely functional, specific, and refined the human body is.

AND WE HAVEN'T EVEN TOUCHED ON THE MOST EXTRAORDINARY ORGAN OF ALL: THE HUMAN BRAIN.

This organ continues to astound researchers. As we say in Puerto Rico, "Está en otra liga", it's in a league of its own. With all that science has discovered about its anatomy and function, the brain remains full of mystery. Experts agree: there is nothing like it.

You might be thinking about artificial intelligence. But don't be fooled, just because a machine can work faster or process more data doesn't make it superior. Computers cannot feel, think, or reason like a human being. Their "intelligence" is built from algorithms and limits imposed by their designers. In contrast, the human brain was designed by our loving Creator and possesses unmatched beauty, adaptability, and depth.

Recognizing this truth leads us to marvel at God's handiwork. He gave humanity a body that is exquisite, intricate, and awe-inspiring. As stewards of this undeserved gift, we are called to care for our health, holistically. That means nurturing the three dimensions that define us: the body, the mind (with its emotions and senses), and the spirit.

These pillars affirm what Scripture tells us: that we were created in the image of God. No less could be expected from a Being so majestic. And while sin has brought disease, aging, and death, we can still hold on to the divine promises of care, protection, and healing when we entrust ourselves to God and cry out to Him in faith. This is the hope we cling to during our brief time on this earth.

Because ultimately, we were created to live forever, a destiny far greater than the limited and destructive offer made to us by the enemy of God. Woven into our spiritual DNA is a divine code, a memory of what once was and what will be again: that we were created to be loved by our Heavenly Father. And He has already done everything necessary to redeem us, to transform and restore us from this mortal state to one of eternal vitality.

Now *that* will be a transformation, of appearance, of health, of freshness, and of total, radiant life.

THE MYTH OF THE FOUNTAIN OF YOUTH AND BEAUTY

When I think back to my elementary school years, I remember studying the Spanish conquest of the New World in our Latin American history class, particularly the period spanning the late 14th and early 15th centuries. Our teacher told us about Juan Ponce de León and his bold quest for the fabled Fountain of Youth. He wasn't alone in this search; many other conquistadors dreamed of finding such a priceless treasure.

If we go even further back in history, the Egyptians stand out as a civilization deeply enamored with beauty and the glow of youth. Many of their customs and ceremonies focused on appearance, aesthetics, and physical health. The famous Queen Cleopatra is said to have enchanted men with a rare and distinctive perfume. A recent CNN article reported the discovery of Egyptian skulls dating back more than 4,000 years that showed evidence of advanced surgical procedures, further illustrating the importance they placed on health and bodily care.

The Greeks were another influential civilization that valued both health and beauty. They introduced a different way of thinking about the physical body, the mind, and what they defined as the soul. During that period, certain philosophies began to portray the body as something sinful or impure, prone to wrongdoing. In contrast, the soul or mind was elevated as the seat of purity and holiness, something to be revered and protected from the body's supposedly corrupt nature.

This dualism created a rupture in how societies perceived and valued beauty, wellness, and health. Philosophers, physicians, and thinkers of the time developed elaborate reasoning and practices aimed at suppressing the body's sinful tendencies, hoping to attain true beauty and wellness by cultivating a pure and refined soul. As a result of these ideas, we have inherited and internalized a fragmented understanding of what it means to be healthy or beautiful, often drawing flawed conclusions about the true nature of human well-being.

Today, science and technology have reached extraordinary heights in their tireless quest for eternal youth and physical perfection. Between myths, legends, historical accounts, and scientific breakthroughs, humanity, consciously or not, remains constantly in search of ways to stay young, strong, and attractive for as long as possible.

Global beauty industry statistics show that in 2022, humanity spent a staggering 504 billion dollars on beauty products. Of that total, 42% went to skincare, 24% to hair care, and 20% to makeup, all in an effort to preserve or enhance outward appearance.

Now consider the spending related to healthcare services aimed at preserving or improving physical health. In the United States alone, the total expenditure for 2022 was an astonishing $4,255.1 billion. Key areas included:

$1,323.9 billion for hospital care, $680.4 billion for other personal healthcare, $633.4 billion for medical services, and $378 billion for prescribed medications.

There is no question that physical appearance and bodily health hold a central and elevated place in human life. Yet this emphasis often creates internal tension and imbalance. It can disrupt the vital harmony, or homeostasis, that should exist between our physical, mental, and spiritual dimensions. Even more concerning is the danger of developing a false dependence on external appearance, which can lead to pride, vanity, and a distorted self-image. We sometimes fall into the trap of assuming that a healthy, attractive exterior reflects full internal health. But clinical evidence often proves otherwise.

This is why we occasionally see individuals with vast wealth and access to the finest treatments projecting an image of perfection, enhanced through expensive products, surgical procedures, and clinical therapies, only for the public to be stunned by revelations of serious illness, mental distress, or even untimely death.

A recent and sobering example is the announcement of Princess Catherine of Wales's cancer diagnosis. No one wishes harm upon her, especially given the assumption that she has access to the best care in the world. Like many, I pray she will overcome this trial. Yet her experience reminds us all of a profound truth: none of us is exempt from the realities of deterioration, illness, aging, or death. The difference lies in how we confront these inevitabilities, how wisely, humbly, and gracefully we walk through them. And the key lies in where God stands in our lives.

THE IMPORTANCE OF HARMONY

In Job 38:7, God reminds Job of the celestial harmony present at the beginning of creation: "When the morning stars sang together, and all the sons of God shouted for joy."

Among those "sons of God" was Lucifer, an exalted angel of dazzling beauty and brilliance. He once rejoiced in praising and serving the Creator. But a discordant note arose from within him, disrupting the harmony of heaven. That note became a strange and selfish melody. No longer did it glorify the divine Composer; instead, it exalted the creature above the Creator. Lucifer became enamored with his own beauty and desired to be equal to God.

This is detailed in Isaiah 14 and Ezekiel 28. The biblical descriptions of this once-beautiful angel at the moment of his creation reveal the love, care, and intentionality of the Creator. Like every sentient being, Lucifer was given free will, the ability to choose a full and harmonious life in God or to separate from his Maker. He had that right. But what Lucifer failed to understand, or perhaps willfully forgot, is what Jesus later affirmed in John 15:5:

"I am the vine; you are the branches. Whoever abides in Me and I in him bears much fruit. For apart from Me you can do nothing."

God, as the supreme Being and Creator, sustains each of His creatures through an invisible golden thread, a spiritual connection through which He imparts the spark of life. This divine spark returns to Him at the end of our days, as Ecclesiastes 12:7 explains:

"The dust returns to the earth as it was, and the spirit returns to God who gave it."

Put simply, to live in the truest sense is to remain connected to the Creator. Only God can sustain genuine harmony in our threefold being, body, mind, and spirit.

We understand that Lucifer and the angels were created differently from humanity, but the source of life is the same: God alone. Tragically, Lucifer rejected inward values such as integrity, humility, gratitude, and love for the Creator. Instead, he bowed to vanity, self-indulgence, envy, and the worship of external beauty. These were foreign and destructive emotions, unknown in the universe until that point, and through the entrance of sin, they were passed down to the human race.

Since the Fall, we've inherited distorted understandings of beauty, purity, and morality. And today, we can observe firsthand the ongoing deterioration of human life. Society often operates in vicious cycles where pride and self-exaltation take precedence over reason, and the truth is obscured: that integrity is far more valuable than mere appearance.

Those who try to uplift moral values are often ridiculed. Those who display altruism, service, and self-sacrifice, who strive toward growth, character, and devotion, are frequently overlooked or dismissed.

A telling example, at least within American society, is the indifference sometimes shown toward the men and women who serve in our armed forces. I had the honor of working for and alongside them for over 30 years. Do we truly understand the depth of sacrifice and commitment that military service entails? The dreams and personal plans put on hold until their duty is fulfilled? The burdens carried by their families, the wounds, visible or invisible, brought home? The trauma, pain, and suffering of those who return without a vital or functional part of their body?

But this is not the point I wish to highlight. On the contrary, I want to speak of their resilience, their fighting spirit in the face of adversity, their determination to overcome obstacles, and their commitment to rehabilitation. These individuals understand that they are far more than outward appearance. They recognize the importance of guarding their minds against discouragement, loneliness, and despair. And they know that, to survive and thrive, they must keep their spiritual compass intact. There is always a true North guiding them.

They are examples of individuals who know how to evaluate themselves honestly. They ask: "*What am I willing to accept as quality of life?*" And more importantly: "*What will I do to rise from where I am?*"

The Bible is full of examples of great men and women who faced immense trials before experiencing the fruits of their faith, perseverance, and devotion.

We read of Hannah, desperate and burdened, who, though barren, prayed for a son and then gave him back to serve God. We see Joseph, sold into slavery, falsely accused, imprisoned, and humiliated, rise to become second only to Pharaoh.

We remember Daniel, taken from his homeland, yet unyielding in his commitment to bodily and spiritual integrity, whose faithful choices earned him a place among the great prophets.

And what of Job? Not only did he lose every material possession, he also lost his children. Yet, in the end, everything was restored to him.

Now consider Sarah, wife of Abraham. In the previous paragraphs, we've reflected on mental capacity, emotional strength, spiritual integrity, and physical beauty. Sarah was a woman who, within the social norms and family dynamics of her time, stood out not only for her beauty but also for her loyalty and perseverance. Twice, she was taken before powerful rulers, Pharaoh and later King Abimelech, each intending to make her their own, captivated by her striking appearance. But both times, divine intervention occurred, and Abraham was honored and enriched by those very kings.

Later, in a moment that defies logic and time, Sarah, at ninety years old, conceived and gave birth to a son. It was as if she had discovered the fountain of youth. Hebrews 11:11 explains:

"By faith Sarah herself also received strength to conceive seed, and she bore a child when she was past the age, because she judged Him faithful who had promised."

All of this affirms a vital truth: human wholeness goes far beyond physical appearance, mental clarity, or spiritual devotion when viewed in isolation. We cannot, and must not, allow one of these areas to dominate the others. The perfection with which our bodies were originally created reminds us that balance is essential.

We begin to perceive this truth when we pause our hectic, hurried lives and humbly observe the deep interconnection between body, mind, and spirit. And, just as importantly, when we recognize and accept that their full development depends not solely on our efforts, but on God's loving hand and His vision of excellence for our lives.

This is the profound importance of what Jesus came to restore in and through us. Not only did He pay the debt of sin, but along with that undeserved gift comes the promise to restore us to our original design. The gift of eternal life includes the restoration of that lost harmony, that true homeostasis that makes our entire being resonate with the rhythm and melody of perfect praise to our Creator.

This restoration becomes real in our lives when we allow our physical body to become a temple of the Holy Spirit, our mind to expand and be elevated as the throne of Christ's fullness, and our spirit to move and vibrate in harmony with the will of the Father. And, of course, we hold the blessed hope that total and complete restoration will be realized when Jesus returns to bring us home to heaven.

I invite you to explore, in the chapters ahead, each of these dimensions separately, learning more about their unique traits and extraordinary capacities.

SELF-EVALUATION AND REFLECTION

1. How do you define beauty, health, vitality, and youth based on your own experience?

2. What lessons do you draw from Sarah's life?

3. How do you understand or perceive the process of aging or maturing? Do you fear it, or do you embrace it with confidence and hope?

4. What is your personal standard for health and beauty?

CHAPTER 3
THE PHYSICAL DIMENSION

GETTING TO KNOW YOUR BODY BETTER

Health, for many people today, has become a luxury. A 2020 study revealed that just over 40% of the global population cannot afford what experts define as a healthy diet. This sobering reality stands in stark contrast to the original divine plan, which envisioned human beings and their environment marked by harmony and excellence. Perhaps it was this excellence that inspired the psalmist to write:

"For You created my inmost being; You knit me together in my mother's womb. I praise You because I am fearfully and wonderfully made; Your works are wonderful, I know that full well." Psalm 139:13–14

Today, many people take for granted, or simply overlook, the extraordinary advantages and capabilities of having a healthy body. But what does it mean to have a healthy body?

According to the Royal Spanish Academy, health refers to the state in which a body can perform all its functions normally. The challenge with such a broad definition is that it assumes a clear understanding of what "normal" is, and what all those bodily functions entail.

Instead of focusing on technical definitions, I invite you to pause and reflect deeply on the marvelous, incomparable properties of the organism we call the human body.

Let's start with the senses: taste, smell, touch, hearing, and sight. Now ask yourself, what is required to activate each one? Take taste, for example. Why does food have flavor? Science has shown that in addition to the taste buds on our tongue, the senses of smell and sight also play a crucial role. These senses work together to help the brain identify and evaluate the flavors we experience. Isn't that fascinating?

Since we're talking about taste, it's only natural to move on to the process of digestion. The Greek physician Hippocrates, considered a father of modern medicine, famously said around 400 BC: "Health begins in the gut." And today's cutting-edge research strongly supports this claim.

You may have heard of the microbiome and microbiota. The microbiome refers to the environment or habitat where microscopic organisms live, and the microbiota refers to the organisms themselves. The discovery of their impact on overall health has revolutionized the field of medicine. Hippocrates was right, more than we ever imagined.

But before diving into digestion and the microbiome in more detail, let's take a step back. What does the basic functioning of a healthy human body look like? Without using complex medical terminology, let me describe some core components of general physical health:

- **Skin** that is smooth, intact, and healthy, with good turgor (firmness), appropriate color, and temperature.

- **A mind and brain alert** to time, place, and person, capable of sending messages and processing information through vast networks of neurons. This includes the central nervous system (with its 12 pairs of cranial nerves) and the peripheral nervous system.

- **An autonomously functioning heart**, supported by a vast network of blood vessels, arteries, veins, and capillaries, that transport oxygen and essential nutrients like vitamins, minerals, enzymes, amino acids, hormones, neurotransmitters, and other chemical components to our cells.

- **Lungs that expand fully and freely** to allow the vital exchange of oxygen and carbon dioxide at the cellular level.

- **A remarkable digestive system** composed of the mouth, teeth, tongue, taste buds, saliva, pharynx, esophagus, stomach, small intestine, large intestine, and anus, all of which work together in highly specialized ways.

- **The liver,** a powerful factory within the body; the gallbladder and pancreas, which produce digestive enzymes, hormones, and essential juices.

- **The kidneys** and urinary bladder, which filter waste and maintain the body's fluid and electrolyte balance.

- **The reproductive organs** of men and women, making the genetic distinction between two biological sexes. These systems are guided by distinctive hormones that influence not only reproduction but also personality, mood, and individual characteristics.

- **A complex muscular system** that provides strength, power, posture, and movement, complemented by the skeletal system, a framework that protects and supports vital organs. These systems work in harmony through tendons, ligaments, and joints to allow motion, balance, and coordinated activity.

- **The immune system**, which constantly defends the body against viruses, bacteria, fungi, and parasites. It includes two major subsystems:

The lymphatic system, which removes waste and toxins from tissues.

The glymphatic system, which performs a similar function in the brain, clearing away neurotoxins and metabolic waste during rest and sleep.

And although each of these systems or organs is, in itself, a complex world of chemical functions and biological processes, they all synchronize and work together in harmony to maintain bodily homeostasis, resulting in a healthy, integrated organism.

Beyond what we've already discussed, we must also consider the body's various stages of development, particularly the aging process and the specific hormonal changes experienced by men and women. Only by doing so can we form a well-rounded and realistic understanding of what health looks like at each stage of life and according to biological sex. As the body grows and matures, its biological needs evolve. This holistic view is the foundation of what is now known as integrative or holistic medicine.

My aim here is to simplify the incredible and intricate processes our body carries out every day. I hope this helps you appreciate the enormous complexity involved in simply having and maintaining a healthy body. Yours and mine function autonomously when we respect them, through proper hygiene, nourishing food, adequate rest, and the care and attention they deserve.

This is why the Bible reminds us in 1 Corinthians 10:31:

"So whether you eat or drink or whatever you do, do it all for the glory of God."

We must never forget that our body is the temple of the Holy Spirit, and we will give an account to God for how we care for and use it.

WE HAVE LOST FOCUS

Let's return to Hippocrates' famous saying: "Health begins in the gut." Now let's explore that idea through a spiritual lens.

When God created the first human couple, He placed them in a perfect setting, a beautiful garden. There, they would engage in meaningful activity through gardening and stewardship of creation. They were surrounded by beauty, fresh air, crystal-clear water, warm sunlight, and an abundance of appropriate, life-giving food, all within a peaceful, stress-free environment.

And what was that original diet? The Bible gives us a clear answer in Genesis 1:29:

"Then God said, 'Behold, I have given you every herb bearing seed, which is on the face of all the earth, and every tree that has fruit with seed in it; to you it shall be for food.'"

The Creator, who understood the design and function of the human body better than anyone, prescribed a plant-based diet: fruits, vegetables, whole grains, legumes, nuts, and monounsaturated fats like olive oil and other natural oils.

Yet over time, humanity has corrupted and strayed from that original, simple dietary plan. Again and again, we've chosen to indulge in a distorted appetite, abusing and overloading our bodies. It's worth noting that sin itself entered the world through eating a forbidden fruit (Genesis 2–3).

On the physical, human level, this drift from the original diet has become more severe with each passing generation. Excessive consumption of meat, gluttony, and the increasing intake of sugary products like desserts, cakes, and processed pastries have taken a toll. Add to that harmful habits like smoking, alcohol use, and other vices, and it's no wonder modern health outcomes are deteriorating.

In developed nations, a significant portion of the diet now consists of highly processed foods. Medical research has repeatedly linked these products to the modern epidemic of obesity, metabolic disorders, and most chronic illnesses. This is where we must remember the unbreakable law of cause and effect. In this light, the old saying proves true: "We are what we eat."

Unfortunately, the news doesn't get better. We must also consider the widespread overuse of prescription medications, over-the-counter drugs, and illicit substances, all of which can harm the body through side effects, long-term damage, or chemical dependency. On top of that, many people suffer from physical inactivity, chronic overwork, lack of rest, and overall intemperance in multiple areas of life. All of this is compounded by mounting stress and debilitating anxiety.

Each of these unhealthy habits chips away at our overall well-being, disrupting the body's delicate balance and interfering with the biological processes essential for maintaining homeostasis. Normally, our bodies are designed to function energetically during the day, and then at night, while we sleep, to repair any damage sustained. This natural repair cycle is essential to prepare us for the day ahead.

But when we neglect proper rest, eat poorly, overwork, and live in constant stress, our biological rhythms are disrupted. As a result, our body's vital functions begin to weaken. And even though disease doesn't arrive overnight, it emerges once the body's natural defenses and regenerative systems become overwhelmed. The body tries to protect us, again and again, but eventually, its resilience is worn down.

In recent years, clinical and scientific research has increasingly pointed to the gut as a central player in this equation. Many of the most pressing health mysteries, including chronic illness, seem to trace back to how we treat our digestive system and the complex ecosystem it hosts.

THE MICROBIOTA AND OUR INTERNAL GARDEN

Since the 19th century, with the rise of modern medicine and continuing into the 21st, scientists have recognized that up to 90% of diseases have a psychosomatic origin. In other words, the mind's influence over the body plays a central role in the development of illness. According to the World Health Organization (WHO), the leading global causes of death include:

- Ischemic heart disease
- Stroke
- Chronic obstructive pulmonary disease (COPD)
- Communicable respiratory infections
- Diabetes
- Cancer

Even in developed countries, hospitals and primary care facilities often lack the resources to meet growing patient demand. One of the core problems in modern medicine is its tendency to treat symptoms, rather than the root causes of disease. This has contributed to what many describe as a pharmaceutical monster, a system in which nearly every symptom has multiple medications that must often be taken for life, without ever producing a true cure.

This dissatisfaction has led to a resurgence in alternative and natural approaches to healing, most notably naturopathy, as well as a newer branch of science known as functional medicine.

Cutting-edge research continues to point to the gut's microbiota as a key factor in both the development and prevention of disease. Scientists have discovered a powerful relationship between the vagus nerve, which connects the brain to the gut, heart, lungs, and other vital organs, and the role of intestinal microbes in sending and receiving critical signals.

This connection has been termed the gut-brain axis. Some even call the gut our "second brain."

But where do these signals originate? Here we arrive at the microbiota, the ecosystem of trillions of microscopic organisms lining the intestinal walls. These include bacteria, viruses, fungi, and parasites, many of which live symbiotically in our bodies. They communicate with one another, control important biological processes, and, when in balance, play a protective role. A healthy microbiota regulates both beneficial and potentially harmful microbes, keeping the system stable.

These microbes exert influence through:

- The vagus nerve,
- The secretion of hormones, metabolites, and neurotransmitters,
- And the production of cytokines, proteins that guide immune and blood cell activity.

Essentially, your gut is a microscopic garden, with its own rich ecosystem. Like any garden, it requires care, balance, and proper nourishment.

It is now understood that 70% to 80% of the immune system resides in the gut. This reinforces the idea that we are what we eat, because what we ingest directly shapes the strength, responsiveness, and balance of the immune system responsible for protecting us against illness.

There is an increasing body of documentaries and research, available on platforms like Netflix, that explore these groundbreaking discoveries about the gut, microbiota, and immune function. Many of them draw connections to what are known as the Blue Zones.

Blue Zones are regions in the world where people regularly live beyond 100 years, enjoying exceptional health, independence from pharmaceuticals, freedom from chronic disease, and a high quality of life.

What do these people have in common?

Research in these areas has identified key lifestyle practices shared across cultures:

- A simple, balanced diet, low in processed foods
- Adequate rest and quality sleep
- Regular exposure to sunlight
- Clean, pure air
- Daily exercise
- A deep sense of purpose and motivation
- Trust in God
- And strong, healthy social relationships

This powerful evidence reminds us how important it is to care for our bodies. What we consume directly affects our vital organs, and our physical well-being influences both mental and spiritual health.

True wholeness depends on all three dimensions: body, mind, and spirit. Proper nutrition and bodily care give strength to the mind and support spiritual clarity. And in turn, a healthy mind and a spirit aligned with God empower us to master our appetites and steward our bodies wisely. These three aspects of being human are inseparable. In the following chapters, we will dive deeper into the mental and spiritual dimensions.

As mentioned earlier, sin entered the world through food. Eve had access to abundant, nourishing options. She was not hungry. Yet, in an act of distrust and disobedience, she chose to eat from the one tree God had warned against.

We see a similar story in Numbers 11. After God miraculously delivered the Israelites from Egypt, providing food (manna), water, and safety in the desert, a rebellion arose among the people, fueled by outsiders and their cravings for meat. The Israelites joined in this rebellion, despite God's faithful provision. In response, God gave them meat for an entire month, but the consequences were severe. As Scripture records, death swept through the camp even as the meat was still between their teeth.

The issue was not the meat itself, but the rejection of God's care and provision. They despised the food God lovingly gave and demanded something else.

Fast forward to the temptation of Jesus in the wilderness. After fasting for 40 days, Satan tempted Him to turn stones into bread. But Jesus responded with these words:

"Man shall not live by bread alone, but by every word that proceeds from the mouth of God." (Matthew 4:4)

Humanity fell through misplaced appetite. But it was through victory over appetite that Jesus began His path toward redeeming us. And in His victory, we find the strength to reclaim our health and restore harmony. If we wish to recover the balance lost through sin and excess, we must begin by restoring our appetite, that is, learning once again to eat with care, intention, and gratitude.

We must:

- Be mindful of what we consume
- Respect the natural laws of the body
- And trust in God's promises

As 3 John 1:2 tells us: "Beloved, I pray that you may prosper in all things and be in health, just as your soul prospers."

This is the divine will for your life: not just spiritual salvation, but physical wholeness and lasting wellness, a harmony that reflects the image of the One who created you.

AVOID THE COMMERCIAL TRAP AND THE MAGIC PILL

If there is one thing you should never doubt, it's this: God, your Heavenly Father, cares deeply about your well-being. Yet the separation caused by sin has led many to forget, or even ignore, the countless promises of care and protection He left us in His Word. One of the clearest demonstrations of this love was through Jesus' healing ministry. Time and time again, He reached out to the suffering, not just to forgive sin but to restore vitality and bring healing to broken bodies.

Then, as now, those responsible for caring for the sick often felt overwhelmed by the destruction caused by disease. In most cases, there was no cure, only treatment aimed at pain relief and symptom management. Today, hospitals and medical systems excel in handling acute and emergency situations: broken bones, heart attacks, surgery, trauma, and more. But when it comes to chronic and preventable diseases, we fall short.

It's tempting to blame private interests, health insurance companies, pharmaceutical giants, but the truth is that the responsibility also lies with us. Many have ignored or discarded the divine method for maintaining health and wholeness. In the book of Deuteronomy, as Moses gave his final words before Israel entered the Promised

Land, he reminded the people that if they obeyed God's commandments and laws, the Lord would keep disease away. None of the illnesses that afflicted the Egyptians would touch them (Deuteronomy 7:15). That promise still echoes today: if we live in alignment with the natural laws of our body, if we honor cause and effect and present ourselves as a "living sacrifice, holy and acceptable to God" (Romans 12:1), we will experience better health.

Even recent findings in functional medicine confirm what Scripture has always taught: many chronic diseases can be prevented, or even reversed, if we are willing to change our habits and embrace a healthier lifestyle.

A SIMPLE BUT TELLING EXAMPLE

My own introduction to healthcare came through my training as a missionary nurse. The university where I studied deeply instilled in us the mission to help restore the image of God in our patients through compassionate care.

In the emergency room, I often encountered patients with severe abdominal pain, vomiting, and rectal bleeding. These were critical cases, marked by dangerously low blood pressure, anemia, electrolyte imbalances, anxiety, and even delirium. We would rush to insert IV lines, administer emergency medications, and place patients on cardiac monitors. If bleeding wasn't controlled quickly, cardiac or respiratory arrest could follow.

In many of these cases, the root cause was alcohol abuse. Once stabilized, the patient would be admitted and, after a few days, discharged, usually with instructions to stop drinking. But for most, this wasn't the first time they'd received the same warning… and sadly, it wouldn't be the last.

As a Christian and missionary nurse, it was hard for me at first to accept that patients have the autonomy to continue harmful behaviors. But I learned that I had to respect their freedom, even when their choices endangered their well-being.

This same pattern repeats itself in the lives of those struggling with chronic diseases: diabetes, hypertension, strokes, arthritis, autoimmune conditions, and more. Conventional medicine often offers medications to manage these conditions, but not to cure them. In fact, most medications are meant to be taken for life, and often come with side effects, some of them severe. Even so, many patients prefer a pill over true lifestyle change.

Meanwhile, there's been an explosion in the use of natural supplements, often without professional guidance. Unfortunately, many of these products are ineffective or even harmful, produced under unsafe conditions, lacking scientific backing, and used without the lifestyle changes necessary to see real improvement.

What Is the Real Solution?

Before we rush to assign blame for our health challenges, we must ask:

- Have I made poor choices?
- Do my habits work against my body?
- Have I overlooked the role of heredity, pollution, or stress?

- Have I failed in my responsibility as a parent or caregiver to teach healthy habits?

There is no magic pill that instantly restores lost health. And while God respects our free will, He longs to guide us back to healing. The enemy wants to trap us in cycles of neglect, indulgence, and disease, but God offers a path of restoration. As it says in Jeremiah 33:6:

"Behold, I will bring health and healing to it; I will heal my people and will let them enjoy abundant peace and security."

SELF-EVALUATION AND REFLECTION

1. How would you describe your physical health at this moment?

2. Are there situations where you have neglected your body, the temple of the Holy Spirit?

3. Looking at the natural remedies mentioned above, can you commit to two or three changes you're willing to make today to recover your physical health?

4. What might be holding you back from going to Jesus and asking Him for help to care for your body and improve your health?

CHAPTER 4

THE MENTAL DIMENSION

THE POWERFUL HUMAN MIND

The world of medicine constantly evolves, adjusting and advancing whenever new evidence calls for it. For this reason, health professionals affirm that modern medicine is rooted in clinical evidence. And it is this very evidence, or the absence of it, that highlights the human brain as an extraordinary organ.

I make this claim because, among all created beings, only humans possess the ability to observe, analyze, express, communicate, change, adapt, choose between alternatives, verify, experiment, feel, interpret, and draw conclusions based on their environment and interactions with it. The human brain carries out these complex processes continuously and simultaneously, something even the most advanced artificial intelligences cannot fully replicate. No matter how sophisticated AI becomes, it remains limited by the parameters and algorithms set by its creators. It cannot feel or reason as you and I do.

This is why the wise Solomon wrote in Proverbs 4:23:

"Above all else, guard your heart, for it is the wellspring of life."

Theologians, considering the context of the chapter, agree that the "heart" in this verse refers to the *mind*. It is in the mind that thoughts are born, where we apply reason, and where we respond to both internal and external stimuli. Most importantly, it is through the mind that we can commune with God. This is the primary reason Proverbs instructs us to guard our mind above all else.

Although sin disrupted our original harmony and connection with God, the Holy Spirit now works through our minds, empowering us, through faith in Christ's sacrifice on the cross, to reconnect with our Creator, the true source of life and power.

This is the very power of the Gospel that Paul proclaims in Romans 1:16:

"For I am not ashamed of the gospel, because it is the power of God that brings salvation to everyone who believes."

If God uses our minds to make salvation effective through faith, then it is essential that we understand the incredible qualities of this remarkable organ.

The average human brain weighs about three pounds (1.4 kilograms) and contains nearly one billion cells known as neurons. These include two main types: standard neurons and glial cells. Each neuron can form over 10,000 connections with other neurons via axons and dendrites. This magnificent organ is the centerpiece of the central nervous system.

Structurally, the brain consists of the cerebral cortex, which is divided into two hemispheres, left and right, connected by the corpus callosum. These hemispheres include the frontal, parietal, temporal, and occipital lobes, each responsible for critical functions such as language, writing, playing music, bodily coordination, memory, visual processing, and more.

Deeper within the brain is the diencephalon, which includes the thalamus, gathering sensory information except for smell, and the hypothalamus, which regulates appetite, temperature, and other vital processes. The midbrain transmits electrical impulses from the brain to the muscles, while the hindbrain, which houses the cerebellum, controls balance and movement. Also in the hindbrain is the medulla oblongata, which regulates essential autonomic functions like breathing, heart rate, and blood pressure.

As you can see, even describing the components of the brain is no simple task. What's more, neuroscientists admit that they still do not fully understand how it works. They recognize that current knowledge is incomplete and must be revised as new discoveries emerge.

Who we are as persons, our reasoning, emotions, thoughts, desires, and decisions, still contains deep mysteries. This is why many in the scientific community regard the brain as the most complex organ in the universe. And for us as Christians, this complexity is yet another testament to the majesty of our Creator!

PRIMARY FUNCTIONS OF THE BRAIN

From the moment we are born, the brain begins collecting information from the world around us. Our senses, taste, smell, sight, touch, and hearing, gradually develop, helping us respond to external stimuli and interact with other human beings. These responses may take the form of laughter, a hug, surprise, pain, indignation, anger, or even rage. Each of these reactions must, in a sense, be shaped and managed so we respond appropriately to the stimulus received. If the brain does not receive sufficient stimulation during the early months and years of life, it loses much of the incredible potential with which it was created.

It is vital to understand that the brain is the control center of our entire being. The electrical impulses it constantly generates, and those it receives from every part of the body, shape who we are and give life to our experiences. Within the brain, our dreams are born. There we form desires and hope; we experience pleasure and fear, joy and sadness. It is in the brain that we learn to love and to forgive, to accept or reject challenges, to rise with the courage that faith gives or to fall into fear or shame. It is also the cradle of altruism, where we serve, protect, and care for others selflessly, or, sadly, where we may lose our way and act out of anger, pride, or contempt. It is where doubt, envy, and discouragement can take root and drive us to actions we later fail to recognize in ourselves.

I vividly remember a moment from my time working in the emergency room. A young man, not yet 25, arrived after a drug overdose in an apparent suicide attempt. I had to act quickly, establishing venous access, connecting him to a cardiac monitor, and performing gastric lavage. It was a chaotic and distressing scene. We had little time to remove the excess drugs from his system and to counteract what had already been absorbed. As I worked, I quietly prayed, asking God to intervene and preserve his life.

By God's grace, we stabilized him. Once the crisis passed, I cleaned the cubicle, wiped away the vomit, and sat beside this young man, someone who, for reasons I could not yet know, believed that ending his life was the only way out. I spoke to him gently, trying to distract and calm his anguished spirit. He was overwhelmed by pain and despair. I took his hand, cooled his forehead with a damp towel, his skin slick with diaphoresis, the medical term for excessive sweating, and stayed with him until he showed signs of peace.

As I returned to the desk to complete my nursing notes, I quietly thanked God for answering my prayer. Just then, the attending physician turned to me and said, with chilling indifference:

"*Why do you waste so much time on someone who clearly doesn't value his life?*"

For a moment, I was speechless. Indignation and anger

surged within me. I took a deep breath and counted to ten, just as I'd been taught in university to manage anger, and I asked God to help me respond wisely. Immediately, the words of Jesus to the Pharisees came to mind. When they criticized Him for spending time with tax collectors and sinners, He replied: "It is not the healthy who need a doctor, but the sick."

And there before me was a young man crying out for help, from the depths of his heart.

I looked the doctor in the eye and said:
"*Every day when I come to work, I ask God to help me serve and reflect His character. And every day, when I go home, the greatest gift God gives me is a clear conscience and the peace that comes from knowing I did my best to help someone. That is what allows me to sleep without guilt or regret. To me, that is true freedom in Christ.*"

I may never know what led that young man to the edge of despair, but I have never forgotten the inner prompting of the Holy Spirit, the voice that warns, protects, and guides us when we begin to lose our way. Jesus is our model. He desires that our lives be fruitful, purposeful, and filled with service to others.

Luke 2:52 tells us:
"And Jesus grew in wisdom and stature, and in favor with God and man."

This is the model for our own growth and for that of our children. From the early years of life, the brain begins to master motor functions, crawling, walking, running, and gradually learns complex skills like language, social behavior, sports, music, arithmetic, science, singing, and dance. All of these are shaped by our environment and the resources available to us.

Researchers agree that abilities and skills are influenced by many factors, including culture, genetics, socioeconomic status, and education. Yet none of these variables on their own can determine a person's success or failure. That's because attributes like personal motivation, resilience, focus, a sense of adventure, willingness to take risks, self-discipline, and trust in God play an equally vital role. These inner qualities are what often determine whether a life is lived with purpose, fulfillment, and peace, or not.

BRAIN CURIOSITIES

Not long ago, scientists believed that we were born with a fixed number of nerve cells, or neurons, and that as we aged, these neurons were gradually lost, leading inevitably to brain degeneration. Today, however, research has proven that while general neurons do not regenerate, glial cells do. Even more remarkably, the brain retains the ability to form new connections between neurons well into our eighties and beyond. This means the brain can remain strong, healthy, and fully capable of continuing its essential role as the control center of the body throughout our lives.

One of the most fascinating discoveries in recent neuroscience is the concept of neuroplasticity, the brain's ability to recover, restructure, and adapt to new circumstances. Neuroplasticity allows for lifelong learning, helps strengthen memory, and gives new hope in the treatment or delay of certain neurological conditions. It may also help explain why some people seem particularly resilient in the face of adversity. Though all of us experience hardship, some are able to persevere, rising above their pain with a spirit of hope and optimism. In this way, the brain functions much like a muscle: if it is not exercised, it gradually loses its strength and flexibility.

In 1996, a team of Italian researchers published a groundbreaking study on mirror neurons, special neurons located primarily in the prefrontal cortex and also in the parietal lobe. What surprised scientists was that these neurons responded not only to our own actions but also to the actions of others. In other words, when we observe someone performing an action, our mirror neurons activate as if we were performing that same action ourselves. This means that a significant part of our learning occurs through imitation, by observing and internalizing the behaviors, emotions, and actions of those around us.

Put simply: we learn what we see. The things we spend time with, the people we surround ourselves with, the habits we practice, and the mentors or leaders we choose to follow all have a powerful influence on how we think and act. This brings new depth to the old saying many of us heard growing up: "Tell me who you walk with, and I will tell you who you are." It becomes clear that we must carefully guard our relationships, our role models, and the sensory input our brains receive each day.

Returning to the subject of mirror neurons, scientists also believe they play a role in empathy, our ability to connect with others emotionally, to understand their feelings, and to share in their experiences. This underlines the importance of what we feed our minds through our senses, because what we take in shapes who we become.

The Bible speaks powerfully to this truth, long before science confirmed it. In 2 Corinthians 3:18, we read:

"And we all, who with unveiled faces contemplate the Lord's glory, are being transformed into his image with ever-increasing glory, which comes from the Lord, who is the Spirit."

Glory to God for such a profound revelation! Scripture has already provided us with the path not only to mental health and renewal but also to spiritual growth. The latter we will explore more fully in the next chapter, on the spiritual dimension. But as it relates to the mind, what matters most is this: by contemplating Jesus, by fixing our eyes on Him and soaking in the beauty of His character, our minds are transformed, day by day. Through faith, we are lifted above the suffering of this world and reshaped into His likeness. And in doing so, we discover true peace and purpose.

ALTERATIONS IN BRAIN FUNCTION

In today's high-stress world, maintaining a healthy mind often feels like a luxury. Numerous physical, social, environmental, and genetic factors can lead to physiological damage in the brain. Such damage may cause alterations, radical shifts, or even complete interruptions in the brain's normal electrical activity, resulting in pathological or destructive behaviors. These behaviors, foreign to normal brain function, can harm not only the individual but also those around them. In many cases, such damage manifests in learning difficulties or motor development delays. Mood swings, difficulty regulating emotions such as anger, and deep emotional struggles like anxiety or depression may occur, sometimes to the point where interest in life itself fades. These more severe cases call for professional intervention, but they also demand something just as vital: a deep well of empathy and support from family and close friends.

One of the most important steps society must take is to stop stigmatizing mental illness. Mental health deserves the same level of attention and care as physical health, just as we emphasized in the previous chapter. This is why we rely on specialized professionals in psychiatry, psychology, and counseling to provide needed care.

Some neurological conditions, such as spinal cord disorders, epilepsy, or nerve-related dysfunction, require the expertise of a neurologist. In cases of physical trauma to the brain, a neurosurgeon is essential. There are also progressive neurodegenerative diseases such as Alzheimer's, Parkinson's, and various forms of dementia. Additionally, emotional and psychological trauma can lead to conditions related to anxiety, alcoholism, isolation, abandonment, violence, abuse, or harassment. These are just as real and important as physical conditions. They represent another vital dimension of the human being, one that deserves equal care and attention in pursuit of holistic health.

According to the World Health Organization (WHO), in 2019 the following data was reported on the most common mental health conditions worldwide:

Condition	% World population	% Kids and teenagers	General Symptoms
Anxiety disorders	301 million	58 million	Fear, excessive worry, and related behavioral disturbances. There are several levels.
Depression	280 million	23 million	Deep sadness, irritability, feeling of emptiness, loss of interest. Most of the day and for at least two weeks.

Condition	% World population	% Kids and teenagers	General Symptoms
Bipolar disorder	40 million	Not reported	Depressive episodes altered with periods of manic symptoms.
Post-traumatic stress disorder	Not reported. High incidence in regions affected by armed conflict.	Not reported	1. Reliving the event. 2. Avoiding recalling the event. 3. Persistent perceptions of a greater threat.
Schizophrenia	24 million	Not reported	Significant perceptual impairment. Changes in behavior, delusions, hallucinations.
Eating disorders	14 million	Almost 3 million	Anorexia nervosa, bulimia nervosa.
Disruptive and antisocial behavior disorders	40 million	Included in the statistics	Persistent behavioral problems.

Condition	% World population	% Kids and teenagers	General Symptoms
Neurodevelopmental disorders; including intellectual development, autism spectrum disorder (ASD), and attention deficit hyperactivity disorder (ADHD).	Unreported figures	Not reported	Considerable difficulties in acquiring and performing specific intellectual, motor, or social functions.

FACTORS THAT INTERFERE WITH MENTAL ACTIVITY

Various factors affect how the brain functions, but none is more harmful than chronic stress. Stress becomes dangerous when the brain is overwhelmed, unable to process information or provide an adequate response, despite the absence of physical injury.

A clear example in today's society is the immense pressure many immigrants face upon arriving in the United States. Often forced to leave their countries for a wide range of reasons, many do so without the time or ability to carefully weigh the pros and cons. Lacking sufficient resources, they enter a new life already burdened by emotional and logistical strain.

This can result in high levels of chronic stress with potentially traumatic consequences. These include the loss of support systems, vulnerability to abuse or exploitation, breakdowns in communication due to language barriers, cultural shock, and overwhelming feelings of powerlessness or inadequacy. Such stressors are not just emotional; they deeply affect brain function and can severely impact mental health.

In situations like these, the mind's ability to face difficulties, discouragement, sadness, pain, and disappointment, and to transform them into opportunities for emotional growth becomes invaluable. The capacity to treasure hope, recover, and start anew, along with cultivating strong self-esteem, is essential for resilience.

As a Latina who has lived in the United States for many years, I vividly remember numerous moments in both my personal and professional life when I faced racism, feelings of inadequacy born from misunderstandings, and cultural shock. Maintaining my confidence in God, investing in personal growth, and learning to understand and accept my new environment were crucial tools that helped me overcome these challenges.

A powerful biblical example of the ravages prolonged stress can cause is found in the life of the prophet Elijah. Chapters 18 and 19 of 1 Kings detail Elijah's struggle against the wicked King Ahab and his ruthless wife, Jezebel. By God's command, Elijah prophesied a drought in Israel, which lasted three years. During this time, King Ahab persecuted Elijah, although God protected him. Influenced by Jezebel's idolatry, the king allowed the worship of Baal, leading the people away from God.

On Mount Carmel, Elijah challenged the 450 prophets of Baal to prove who the true God was, demanding the people choose whom they would serve: God or Baal. When God sent fire to consume Elijah's offering, His supremacy was vindicated. Elijah then ordered the execution of the false prophets and prayed for rain to end the drought, a prayer God answered. But when Jezebel heard what had happened, she vowed to kill Elijah. Hearing this, Elijah fled in fear.

Here is where the story turns deeply human. In 1 Kings 19:4 we read: "*He went a day's journey into the wilderness, came and sat under a juniper tree, and prayed that he might die. 'It is enough; now, O Lord, take away my life, for I am no better than my fathers.'*"

After tremendous tension, overwhelming responsibility, and the threat of death, Elijah succumbed to despair. Stress, exhaustion, mockery, and contempt weighed him down until he wished for death. James 5:17 reminds us:

"Elijah was a man with a nature like ours."

Elijah, the great prophet, was human, subject to the same passions and struggles we all face.

What pushes our minds into such tension? Here are some common factors:

- Stress
- Lack of rest
- Poor nutrition
- Excessive stimulation
- Overwork
- Environmental toxins such as radiation, heavy metals, pollution, and pesticides
- Alcohol, smoking, drug use, and certain medications.

I want to highlight excessive stimulation, which has become one of the 21st century's epidemics. Technology, computers, phones, televisions, and smart devices, has revolutionized life but also created chaos, especially for children and youth. Psychiatrist Dr. Marian Rojas Estapé and other experts note that excessive technology use impacts developing brains similarly to how dopamine affects attention, mood, behavior, learning, motivation, and the sensation of pleasure and reward.

For example, activities like video games or smartphone use release dopamine, making users feel pleasure and driving them to spend increasing amounts of time to achieve that reward. This creates a dependency similar to addiction seen with drugs like cocaine, where more of the stimulus is needed to reach the same level of satisfaction. You may have seen videos of young people who, when separated from their devices, lose control, becoming aggressive, using offensive language, or even resorting to violence.

Surprisingly, many technology creators impose strict limits on their own children's use, or ban it altogether during childhood, because they recognize these effects on developing brains.

Given all this, it is essential for us to make a deliberate, 180-degree change in how we live, how we interact with technology, our environment, social relationships, and learning. As a society, we must do better at protecting our children from harmful external influences. Children with healthy minds are far more likely to become the great thinkers, inventors, lawyers, doctors, artists, legislators, teachers, entrepreneurs, and leaders of the future.

So, with all this in mind, what practical steps can we take to protect our minds and keep our brains healthy?

PRACTICAL STEPS TO PROTECT AND STRENGTHEN THE MIND

1. Establish a Consistent Sleep Pattern

Go to bed before 10 p.m. and wake up early at a fixed time each day. Adults need at least seven hours of sleep; teenagers require eight to ten hours, and school-aged children need nine to twelve hours. Create a dark, restful environment by turning off or dimming all lights and, above all, switching off electronic devices well before bedtime.

2. Maintain a Balanced, Plant-Based Diet

Focus on whole, unprocessed foods with plenty of fruits and vegetables. Avoid cold cuts, canned foods, sweets, sodas, and artificial colorings, which can negatively affect brain health.

3. Practice "Earthing" or Grounding

Walking barefoot on natural grass or soil helps reduce inflammation and muscle pain, prevents heart disease, decreases anxiety, and improves mood by reconnecting with the earth's natural energy.

4. Minimize Exposure to Environmental Pollutants

Avoid harsh chemical disinfectants, pesticides, harmful ingredients in cosmetics, radiation, and other environmental toxins that can impair brain function.

5. Listen to Relaxing Music

There are many sources online or via smartphone apps where you can find calming music. Avoid constant noise and sudden jarring sounds to promote mental calmness.

6. Engage in Physical Exercise

Incorporate activities that raise your heart rate and activate muscles, such as walking, yoga, Pilates, or deep breathing exercises. Physical movement benefits both body and brain.

7. Consider Red Light or Frequency Therapy

Emerging studies suggest these therapies can improve mood and reduce stress by exposing the body to specific wavelengths of light, including near-infrared. Always consult a qualified professional before trying this.

8. Challenge Your Mind with Focused Mental Exercises

Activities requiring concentration and focus help maintain and improve cognitive function.

9. Meditate on Jesus and His Sacrifice

There is nothing healthier for the mind than daily meditation on the beautiful character of Jesus and His atoning sacrifice. As 2 Corinthians 3:18 reminds us, contemplating the Lord's glory transforms us into His image.

10. Use Supplements Wisely

Magnesium, zinc, vitamin C, vitamin B12, and the full B-complex vitamins are essential for brain health. While a healthy, varied diet increases your chances of obtaining these nutrients naturally, supplements can help when necessary. Always consult a physician before beginning any supplementation, especially if you have chronic health conditions. Naturopaths and functional medicine practitioners can also offer valuable guidance for those seeking natural lifestyle changes.

Remember Hippocrates' timeless wisdom: "Let food be thy medicine, and medicine be thy food." This principle is regaining importance as a foundational approach in modern health care.

Finally, nothing is more powerful than trust in God.

CHOOSING THE BETTER PART

What determines the path we choose in life? What shapes the inner motives behind our actions? While science attempts to answer these questions, God not only provides the answer but also the source of our strength and true value as human beings.

Two verses come to mind: Philippians 4:13 says "I can do all things through Christ who strengthens me."

And John 15:5 reminds us "I am the vine, you are the branches. He who abides in Me, and I in him, bears much fruit; for without Me you can do nothing."

Often, I wonder why we let the enemy discourage us and surround us with sickness, anguish, and sorrow, when Jesus has already provided the solution to all our problems. The answer lies in walking through life with Jesus by our side. He fights our battles, helps us conquer new heights, and brings joy and peace to our hearts.

In Matthew 11:28–30, Jesus, understanding the heavy and unjust burdens of this sinful world, offers us comfort:

"Come to Me, all you who labor and are heavy laden, and I will give you rest. Take My yoke upon you and learn from Me, for I am gentle and lowly in heart, and you will find rest for your souls. For My yoke is easy and My burden is light."

Do you want to find rest for your weary life, burdened by stress, guilt, violence, discouragement, or despair?

Christ is the answer.

He longs to enter your life, cleanse your weary soul, and restore the hope, energy, and strength you need to triumph. Jesus will bring mental harmony, elevate your mind to new heights, organize your thoughts, and nurture new neural connections that lead to optimism, joy, and the peace you need to fully enjoy life and strengthen your relationships. Give Jesus the opportunity, I promise your life will change for the better.

When Martha complained to Jesus that her sister Mary wasn't helping with the household chores, Jesus' response teaches us an important lesson. He said:

"Martha, Martha, you are worried and troubled about many things. But one thing is needed, and Mary has chosen that good part, which will not be taken away from her." (Luke 10:41–42)

What Martha did wasn't wrong, we all must tend to the responsibilities of life. But we must never let these take first place in our minds. Christ must always be our center, our motivation; everything else is secondary.

Here lies the solution for your mind weighed down by life's burdens: to sit at Jesus' feet, like Mary, who chose the better part.

SELF-EVALUATION AND REFLECTION

1. Take an honest look within. How different are you now from who you were two years ago? Has the change been for better or worse?

2. Can you identify one or two current sources of stress in your life?

3. Looking in the mirror, can you name three positive qualities of your character?

4. Who are the people that motivate you, those you'd like to emulate?

5. How are you feeding your mirror neurons? Consider the places you go, the people you spend time with, and what you read or listen to.

6. If you had a personal appointment with Jesus today, what areas of your life would you invite Him to change?

CHAPTER 5

THE SPIRITUAL DIMENSION

CREATED BY A GOD OF LOVE

At last, we arrive at the most profound and vital dimension of human existence: the spiritual. Whether we acknowledge it or not, we are spiritual beings. We were created by an infinite Being who moves and operates in realms, and likely dimensions, we cannot perceive or comprehend. Recognizing that we are the handiwork of a majestic Creator who expresses, shares, and manifests love in all He does is the foundation of the everlasting gospel.

Before the foundation of the world, before the universe as we know it came into existence, God was, is, and will always be. He is the very essence who guides, sustains, and breathes life into all creation. In Him all living things subsist, develop, and find meaning, on Earth, among the angelic hosts, and in heavenly realms we have yet to see. As Paul says in 1 Corinthians 4:9, we are "a spectacle to the whole universe," an open book to beings beyond our sight.

What, then, does this understanding of our divine origin mean for us? How does the spiritual dimension manifest in our lives? And what value or benefit do we gain from embracing our spiritual nature?

These are the questions we will explore in this chapter, guided by the light of God's Word. What is the essence of my existence? What is my purpose in life? Can an infinite God truly care about me? Why is this world filled with disease and suffering? Why must we die, and what lies beyond death?

When faced with such profound questions, the only safe course is to seek answers in the Word of God.

Nothing pleases God more than when His children come to Him for guidance. He promises in Jeremiah 33:3:

"Call to Me, and I will answer you, and show you great and hidden things that you do not know."

God's Word is our spiritual compass, an unfailing guide that lights our path and directs us safely through the twisted and dangerous roads of life. And as Solomon declares in Proverbs 9:10:

"The fear of the Lord is the beginning of wisdom, and the knowledge of the Holy One is insight."

So let us open our hearts to Scripture and allow God's Word to lead us into a deeper understanding of our spiritual nature, who we are, why we're here, and where we're going.

IN THE BEGINNING

Two powerful currents attempt to explain the origin of humanity and the world we inhabit. Some thinkers even try to merge them, but without success. The first is Creationism, rooted in the biblical account of a beautiful, divinely orchestrated creation completed in one week, followed by a cataclysm: the universal Flood, which radically altered that original world and marked a new, though limited, beginning.

The second is the Big Bang theory, which includes the theory of evolution to explain the origin of species over millions of years. It also incorporates catastrophic events, such as the supposed extinction of the dinosaurs by a meteor or comet. Yet neither view, Creation nor evolution, offers what the scientific community calls evidence-based proof. That is, neither model can reproduce the events in question under controlled conditions. Many inferences are made, especially in the Big Bang model, and these inferences are often presented as fact, though they are not verifiable.

None of us were there millions of years ago to witness the events or understand the exact conditions under which they occurred. This means that faith is required to accept either explanation. Ultimately, it comes down to choosing to believe what God says in His Word, or to accept the enemy's proposal that we are nothing more than the product of chance. We stand either with God or with Satan.

And that brings us to the heart of the matter: the Bible itself defines faith in Hebrews 11:1,

"Now faith is the substance of things hoped for, the evidence of things not seen."

If the key to understanding our origin is faith, why not place that faith in the One who longs to show you, in a thousand ways, how deeply He loves you?

The creation of humanity was an act of love by a God who delights in friendship, companionship, and personal connection. This infinite God created perfect beings and placed them in a perfect world. But as we've already seen, His enemy appeared and distorted everything by introducing sin, and with it, human misery. The very first thing sin brought into this newly created world was fear. And ever since, generation after generation has fallen into that same trap, fear that cripples, paralyzes, and prevents us from truly seeing and knowing the beauty of God's character and His infinite love.

Fear is foreign to the nature of God and an enemy of joy, peace, and harmony. Satan uses fear to destroy us and rob us of our God-given potential. But God, being love's very essence, draws near to us, again and again, to free us from fear's dreadful grip. As 1 John 4:18 assures us:

"There is no fear in love, but perfect love casts out fear, because fear involves torment. He who fears has not been perfected in love."

The remedy is simple: when we are in Christ, walking with Him and keeping His Word, we are perfected in love. God removes every trace of fear. We begin to enjoy a full life, one marked by freedom from worry, distrust, insecurity, shame, and pride. The spirit of service for which we were created is renewed within us. Life becomes pleasant, meaningful, and vibrant, and within us, a golden link forms that harmonizes our three internal dimensions. Our body, mind, and spirit come into alignment.

Our mind is elevated, allowing us to grasp complex truths and cultivate virtues like motivation, empathy, integrity, humility, and a generous spirit. And this is precisely where we understand the beauty and power of God's law of love. As Jesus said in Matthew 22:37–40:

"'You shall love the Lord your God with all your heart, with all your soul, and with all your mind.' This is the first and greatest commandment. And the second is like it: 'You shall love your neighbor as yourself.' On these two commandments hang all the Law and the Prophets."

When we connect with God through our spiritual dimension, the mind, through mirror neurons, begins to reflect the beauty of the divine character. This places us back in harmony with God, with ourselves, with others, and with the world around us.

You may be wondering, "But what about the physical dimension?"

I'm happy to say that it is not overlooked. On the contrary, Scripture is very clear on this point. Recall

1 Corinthians 10:31, which we explored in the previous chapter: "So whether you eat or drink, or whatever you do, do all to the glory of God."

That verse is further strengthened by others that broaden its context. Philippians 4:13 tells us: "I can do all things through Christ who strengthens me."

And in John 15:5, Jesus Himself says:

"I am the vine, you are the branches. Whoever abides in Me and I in him bears much fruit, for apart from Me you can do nothing."

We see, then, that God approaches us through our spiritual dimension, ennobles the mind, and calls it to higher, nobler things. And in turn, the mind governs the body with wisdom and reverence, prompting us to care for it, honor it, and nourish it with every natural resource God has provided. This is what I call the synchronization of our dimensions, body, mind, and spirit, aligning with the ultimate source of life and power. When that alignment takes place, we begin to live with excellence, vitality, and resilience, able to face sorrow, pain, and obstacles with hope, courage, joy, and peace.

LIVING SPIRITUALLY

By accepting Jesus' offer of love, we become new creatures, true representatives and children of the King of the universe. Paul writes in 2 Corinthians 5:17:

"Therefore, if anyone is in Christ, he is a new creation; old things have passed away; behold, all things have become new."

What a beautiful promise!

This is why our spiritual dimension must take the lead. Once we find our center in Christ, we can confidently live the promise of Matthew 6:33:

"But seek first the kingdom of God and His righteousness, and all these things shall be added to you."

Our minds begin to dwell on whatever is true, honorable, just, pure, lovely, commendable, excellent, and praiseworthy (Philippians 4:8). As a result, we cultivate strong, healthy, and creative minds, free from anguish, resentment, guilt, fear, or sorrow.

Physically, our bodies begin to thrive. Every part of us, muscles, bones, circulation, skin, heart, lungs, liver, and brain, begins to function in sync with what our renewed mind envisions and our spirit builds. When we care for our bodies as the temple of the Holy Spirit, we begin to overcome the ailments and weaknesses that result from a diseased condition.

We come to appreciate the value of good hygiene, proper rest, sunlight, pure air and water, a low-stress environment, a healthy balanced diet, and most importantly, trust in God. Our entire being begins to vibrate in harmony under the direction of the Divine Creator, who not only made us but knows exactly what we need to live well. He alone can restore the harmony lost through sin.

It is impossible to achieve this by our own strength. But knowing our weakness, Jesus says in Matthew 11:28–30:

"Come to Me, all you who labor and are heavy laden, and I will give you rest... For My yoke is easy and My burden is light."

And in John 14:27, He adds:
"Peace I leave with you; My peace I give to you. I do not give to you as the world gives. Do not let your hearts be troubled, neither let them be afraid."

Amid a world frantic with haste, power struggles, addictions, immorality, greed, disease, crime, and death, God offers us rest and peace.

DIVINE PEACE AND REST BRING FREEDOM

As I mentioned earlier, I appreciate how Spanish psychiatrist Marian Rojas Estapé encourages her readers and patients to seek out alternatives that nurture joy, motivation, and emotional balance. She speaks often about the importance of setting goals, embracing new challenges, and surrounding ourselves with what she calls **"vitamin people", **those who uplift and energize you. She promotes nurturing the inner self to find balance and peace, noting that a little over 90% of the things we worry about never actually happen.

Just think of the weight of that statistic. Once again, we return to the truth: fear paralyzes us. Worry, doubt, discouragement, sadness, insecurity, anxiety, even depression, are all expressions of the same root cause: fear.
But perfect love casts out fear. The antidote is the pure, healing love of our Heavenly Father.

And how does love shatter fear's grip? By replacing

its toxic effects with inner peace and divine rest, rest that rejuvenates the body, soothes the nervous system, and breathes new life into a weary soul. This is what it means to walk in the Spirit.

Paul gives this advice in Galatians 5:16–17:

"Walk in the Spirit, and you shall not fulfill the lust of the flesh. For the flesh desires what is contrary to the Spirit, and the Spirit what is contrary to the flesh; they are opposed to each other, so that you do not do what you want."

Paul teaches us that walking with Jesus is how we experience true freedom. And having established the path to freedom, we now explore how to receive the peace and rest God longs to give us.

Interestingly, the first time the word rest appears in Scripture is in Genesis. After the glorious work of creation, Genesis 2:1–3 tells us:

"Thus, the heavens and the earth were finished, and all the host of them. And on the seventh day God ended His work which He had done, and He rested on the seventh day from all His work which He had done. Then God blessed the seventh day and sanctified it, because in it He rested from all His work which God had created and made."

God did not need rest; sin had not yet touched the earth. But in His wisdom, He established the importance of pausing every seven days to meditate, contemplate, and rest from ordinary labors.

As we've seen throughout this book, our being includes three interwoven dimensions, physical, mental, and spiritual. And in the spiritual dimension, embracing a weekly day of rest provides the nutrients and renewal that sustain the other two. It is through this rest that we achieve the internal harmony we long for, enabling us to thrive in every area of life.

Think of the Blue Zones, those regions of the world where people live longer and more vibrant lives. One of the common denominators? Rest. And not just physical rest, but a spiritual rhythm rooted in divine connection.

The psalmist captures this truth in Psalm 4:8:

"In peace I will both lie down and sleep, for You alone, O Lord, make me dwell in safety."

And Paul echoes it in Hebrews 4:9–10, saying:

"There remains therefore a rest for the people of God. For the one who has entered His rest has also rested from his works, just as God did from His."

"So then, there remains a Sabbath rest for the people of God; for whoever has entered God's rest has also rested from his works as God did from His."

Resting in the Lord Brings Peace to Mind and Heart

HOW TO FACE PAIN, ILLNESS, AND DEATH

We cannot conclude this chapter without acknowledging the reality of the world we live in, a world burdened by sin, marked by misery, discouragement, stress, disease, suffering, and death. These are not the works of a loving God but of His enemy, Satan, who seeks to destroy humanity by severing us from our source of life and power. Worse yet, he convinces many that God is the cause of every disaster. As a result, people distance themselves from the very One who alone can bring healing and peace, especially in moments of grief, terminal illness, or tragic loss.

But in moments like these, we must ask ourselves a few honest questions:

- Are we truly connected to God through our spiritual dimension? If not, why do we expect Him to intervene on our behalf when we've had no desire to walk with Him?
- Are we reaping what we have sown? Have we neglected the divine principles meant for our well-being?
- Have we placed something, or someone, in the place of God? What or who holds first place in our lives?

Even when our answers appear to be in good standing, one truth remains paramount: God's supremacy and love are unchanging. He is still God. His character remains pure, just, and merciful, even amid human suffering. Our circumstances do not define His nature.

Faith, then, becomes the decisive factor in shaping a character wholly entrusted to God. Faith is what identifies His regenerating power at work in us.

A young mother dies of cancer, leaving behind three beautiful daughters, but in faith, she trusts that the next time she opens her eyes will be on resurrection morning, when she will embrace them again. The grieving parents of a girl who was violated and murdered forgive the offender, understanding that only forgiveness brings spiritual and mental healing. Families rebuild after natural disasters destroy everything they owned, rising with new motivation. These examples are not rare, they are the testimony of people who choose to cling to God despite their pain.

Like Job, they can say:
"I know that my Redeemer lives,
and that He shall stand at last upon the earth;
and after my skin has been destroyed,
yet in my flesh I shall see God,
whom I myself shall behold, and my eyes shall see, not another." (Job 19:25–27)

God does not cause suffering, but He alone can rescue us, unharmed and victorious, from the enemy's grasp.

TRUST IN GOD

As a nurse, I had the honor, and the heartbreak, of accompanying many people in their final moments. I witnessed those who once denied God, even professed atheism, undergo a profound transformation when faced with death. Some were connected to machines; others gasped their last breath. Many tried to reconcile family conflicts they had long avoided. But most of all, they felt alone. Loneliness is one of the greatest afflictions of our generation. Despite living in a hyperconnected world, global statistics show that one in four people, 25% of the population, feels lonely.

But this does not need to be your reality, or mine.

God created us not only for human companionship, but for something deeper: an intimate relationship with Himself. He longs to provide love, sustenance, motivation, hope, joy, peace, comfort, healing, strength, everything our soul needs. Every good gift comes from Him. As Jesus says in Matthew 7:11:

"If you, then, who are evil, know how to give good gifts to your children, how much more will your Father in heaven give good things to those who ask Him!"

Your spiritual dimension holds the key to a full, abundant life. When you are connected to the Source, a world of possibilities opens before you, a life rooted not in fear, but in faith. Just look at the biblical heroes who lived by this truth: Daniel, Esther, Joseph, David, Deborah, Abraham, Moses, Paul, Mary, John, and so many more.

Psalm 46:1–2 reminds us:
"God is our refuge and strength,
a very present help in trouble.
Therefore we will not fear,
though the earth be removed
and the mountains be carried into the midst of the sea."

Scripture overflows with assurances of divine protection and peace, written to silence the voice of fear and remind us who we belong to. As Isaiah 41:10 says:

"Fear not, for I am with you;
be not dismayed, for I am your God.
I will strengthen you; I will help you;

I will uphold you with My righteous right hand."

Do you grasp the beauty of those words? Do not fear. Place your trust in God, and He will act. Invite His Spirit into your heart, and you will witness His plan unfold in ways you never imagined.

SELF-EVALUATION AND REFLECTION

1. Reflect on the last five years. What events or personal changes have helped you grow spiritually?

2. Considering your spiritual walk, what image do you currently hold of God and His character?

3. Have you ever felt trapped by pain, fear, discouragement, or loneliness? Are you facing any of these now?

4. If you had to decide today, would you choose to trust God with those burdens? Would you allow Him to free you from fear and guide your life with His peace?

CHAPTER 6

LIVING TO THRIVE

BREAKING OLD MOLDS

We have now reached the end of our journey exploring the three human dimensions and the importance of embracing them holistically as the first step toward restoration and full health. Often, it is difficult to pause in the frantic rush of daily life and identify what is truly holding us back from growing, progressing, and achieving new dreams and goals. Yet we've also seen that change is not impossible. When we focus on the principles and practices that genuinely make a difference, the path becomes not only manageable but also more joyful and fulfilling.

Especially when we understand that true passion, strength, renewal, and growth come from total surrender to the One who created and formed us for greatness. It is

Christ in us, the hope of glory—our center and potter—who, with His divine hands, reshapes us into His image and likeness.

God wants to help you, heal you, and, most importantly, save you—so that you may live eternally with Him. For this to become a reality, it's essential to observe the natural laws that God, in His infinite wisdom, established for our benefit. We've reflected on several of these, such as:

- **The law of cause and effec**t, which reminds us that our choices—whether wise or foolish—inevitably shape our outcomes.

- **The law of contemplation**, which teaches that what we behold and invest our attention in will, through the activity of mirror neurons, influence and shape our thoughts, emotions, and behavior. Our senses were given as sentinels—to guard the gates of the mind and protect us from evil.

Ultimately, we are called to cast off and break free from modern molds that confine and diminish us—sedentary routines, self-indulgence, gluttony, chronic stress, trivial pursuits, and the addictive, shallow excitement that masquerades as happiness but robs us of true rest. These patterns do not enrich your life; they open the door to illness and premature decline.

The path to a new life—a healthy, meaningful, and prosperous life—begins with a firm, intentional decision to care for your body, mind, and spirit.

HOLISTIC HEALTH

To achieve optimal well-being, you must take an integrative, whole-person approach. A strong mind and spirit will shield you against the fluctuations of the physical world—be they inherited vulnerabilities or environmental challenges beyond your control. They also empower you to overcome obstacles that may threaten your health.

As we review what we've learned from each dimension of human life, let us reclaim the key insights that can lead to lasting, positive change.

PHYSICAL DIMENSION

I do not know your current physical condition, your age, or whether you have been diagnosed with any illnesses or suffer from ailments that affect your health. Nor can I know your environment, resources, or level of knowledge about health. But what I can assure you with certainty is that there are several practices you can adopt or implement to begin enjoying better physical health. Here are a few:

Do not forget the eight natural remedies; these form the foundation for the proper functioning of your body. They are the deep roots that sustain your physical health and support the other two dimensions. The strategy is simple: care for your body from a young age to prevent illness and, whenever possible, halt premature deterioration so that you may reach old age with vitality, stability, and health.

It is essential to care for and strengthen the microbiota in your gut. This is where 70 to 80 percent of your immune system resides—the largest and most complex defense system known to humanity. Called the "second brain," your gut produces 90% of the communication your brain receives through the vagus nerve and neurotransmitters.

Learn about natural remedies found in plants and foods. Many plants and spices are well known for their healing properties—aloe vera, mint, dandelion, mushrooms, turmeric, lemon, cinnamon, garlic, onion, carrot, pepper, ginger, cloves, star anise, avocado, beets, and many others. God has given us the medicine our bodies need in nature to preserve health.

Caring for your physical body requires dedication and attention. Neglecting it will lead to disease and premature aging.

Movement is crucial. A body in motion maintains health. As the saying goes: if you don't use it, you lose it. Exercising your muscles, tendons, and joints helps keep your bones and organs healthy.

Remember, we care for our bodies not to show off a slim or healthy figure, but because we understand and accept that our body is a temple of the Holy Spirit. We belong to God by creation and redemption.

MENTAL DIMENSION

It is essential to understand that neglecting your physical dimension unfortunately places your mind at a disadvantage, as chronic and untreated physical ailments tend to affect mental health. Perhaps you've carried burdens for a long time—past traumas, discouragement, or depressive tendencies—but there is no reason to despair. Recovery is always possible.

We have learned that our brain possesses elasticity, enabling it to create new neural pathways and adaptive patterns to meet ongoing challenges. Neuroplasticity is the brain's, specifically neurons', ability to reorganize into new sequences, allowing continuous learning and growth.

Here are some key points to strengthen your mental health:

- **Rest is fundamental.** The brain needs it daily to perform essential repair functions.
- **Managing chronic stress must be a priority.** The mind cannot function properly when the brain is in a constant state of emergency.
- **Self-care** should include activities that promote relaxation, such as meditation, grounding, hot and cold-water baths, walking in nature, and deep, synchronized breathing.
- **Maintaining a positive attitude** and speaking words of affirmation send your brain the right signals to cultivate motivation.

We care for our mind because it is the pathway God uses to communicate with us, especially through the prefrontal cortex—the seat of decision-making.

SPIRITUAL DIMENSION

The spiritual dimension exists within each of us but is activated through the function of our prefrontal cortex—the part of our brain responsible for conscious decision-making—when we understand that we were created by a loving God who cares for and protects us. The benefits of awakening this spiritual dimension under the guidance of the Holy Spirit are evident both in our inner being and in how we relate to others. Here's why:

- Through our spiritual dimension, we come to truly understand who we are, where we come from, and where we are going. This understanding gives our lives deep meaning.

- Our divine Creator is a God of love who delights in having a personal relationship with His children.

- God's guidance, care, and protection bring countless blessings to our daily lives.

- Placing our trust in God is the key to success both in this life and the life to come.

- A healthy spiritual dimension resting in God as our Creator and Sustainer opens the way to integration and health in both the physical and mental dimensions.

LIVING IN FULLNESS AND VICTORY

Throughout the centuries, human beings have waged an ongoing internal struggle to achieve personal growth. Some attain great victories in the physical realm, where a healthy and strong body gives them advantage. Others rely on sharp minds—great strategists who envision success from the start and lead others to triumph. Many succeed by leveraging physical beauty in various professions, shaped by societal standards of what defines true beauty.

Here we observe a human tendency to be swayed by collective influence, which often turns out to be superficial.

There are also those who try to control others by claiming superior spiritual insight or elevated states, professing to understand esoteric truths more clearly than others.

As you can see, most people focus on one dimension—physical, mental, or spiritual—to stand out or rise above the rest. But the real question is: where do we find the happiness, fulfillment, or satisfaction we long for? How do human beings reach their highest potential?

Reflecting on your own experiences, you may realize that true fullness of life—the happiness and personal satisfaction we seek—is not found in power, outward beauty, wealth, education, resources, or mental strength alone.

To achieve genuine inner harmony, holistic health, and mental peace, we must find balance among our physical, mental, and spiritual dimensions. This balance is only possible when the vital energy that drives and supports these three dimensions is integrated into our way of living.

You may ask, what is this vital energy? The answer is love.

I mean pure, selfless, sacrificial love—the kind the apostle Paul describes in 1 Corinthians 13:4–7:

"Love is patient, love is kind. It does not envy, it does not boast, it is not proud. It does not dishonor others, it is not self-seeking, it is not easily angered, it keeps no record of wrongs. Love does not delight in evil but rejoices with the truth. It always protects, always trusts, always hopes, always perseveres."

This kind of love is found only in the precious character of Christ Jesus. Out of love, He left the heavenly courts and clothed Himself in humanity to die for you and me on a cruel and miserable cross—all for love, so that we might be rescued from sin and granted eternal life. He lived and suffered among humankind so we would understand that only love has the power to overcome every obstacle. Only love enlightens the mind, strengthens the spirit, and brings vitality to the body.

Have you ever heard of a purported letter written by the famous German physicist Albert Einstein to his daughter Lieserl, where he speaks of love as the true universal force? While many articles today claim the letter is a forgery, some of its sentiments ring true. The letter describes love as an immensely powerful force science has yet to explain. It says love is light because it illuminates both the giver and the receiver. It creates attraction among people, multiplies the best within us, and prevents humanity from destroying itself through selfishness. Love reveals and unveils; for love, we live and die. It also states that we fear love because it is the only energy human beings have not learned to control at will.

Because of all these qualities of love, we can understand the words of Matthew 5:48:

"Be perfect, therefore, as your heavenly Father is perfect."

These words were spoken by Jesus in His famous Sermon on the Mount, specifically in the closing section where He calls us to love our enemies.

When the Bible speaks of perfection, it is not referring to being free from mistakes or failures, but rather to loving as Jesus did—to grow spiritually as we allow the Lord to fill us with His love, so we may live that love within ourselves and share it with those around us. In this sense, we walk the path of sanctification. We love because He first loved us. We receive pure and true love from the original source so that we can spread it in this world, where pain, disease, hatred, evil, sadness, and injustice prevail. Only love can overcome these negative forces.

As we progress in this direction, our dimensions become balanced. We are transformed day by day until we achieve victory and divine excellence in our lives. As Ephesians 4:13 says:

"Until we all reach unity in the faith and in the knowledge of the Son of God and become mature, attaining to the whole measure of the fullness of Christ."

Do you want to achieve victory in your life? Do you wish to thrive in everything you set out to do despite the challenges you face? Would you like to gain peace of mind, physical health, and live confidently amid the chaos of this world? If you are willing to be an overcomer and reclaim your state of holistic health, simply give your heart and soul to Jesus, and He will take care of the rest.

If you have tried everything and still haven't seen the results you long for, give Jesus the chance. In His Word, He promises:

"But now, this is what the Lord says—He who created you, Jacob, He who formed you, Israel: 'Do not fear, for I have redeemed you; I have summoned you by name; you

are mine. When you pass through the waters, I will be with you; and when you pass through the rivers, they will not sweep over you. When you walk through the fire, you will not be burned; the flames will not set you ablaze.'"
(Isaiah 43:1–2)

God wants to heal the illness that burdens you. He wants your mind to find the peace it longs for, and your spirit to be revived and refreshed by the beauty and perfection of His presence.

Make today your day of victory. Remember, you were created for greatness!

About the Author

Laura R. Fernández, MPH, BSN, RN, is a respected health professional and passionate advocate of the Christian faith. Born on the beautiful island of Puerto Rico as the youngest of eight siblings in a Christian home, Laura has dedicated over 36 years to the care and well-being of patients in pediatrics, intensive care, and healthcare administration.

She holds a Bachelor of Science in Nursing from Antillean Adventist University and a Master of Public Health from Penn State University, specializing in bioterrorism and disaster management. Laura now combines her health expertise with her love for teaching and the gospel by writing, delivering motivational talks, and preaching.

Happily married and mother to two daughters, Laura enjoys serving her church community and has participated in missionary projects worldwide—from Peru to Kenya.

From a young age, Laura demonstrated a natural gift for communication, excelling in oral and written expression. This gift inspired her commitment to writing and serving others.

She collaborated on an interactive coloring book about the Plan of Redemption used as an evangelistic tool through

Prints of Hope International. In 2012, she published her first book, My Name is Woman: The Plan of Salvation from a Female Perspective, addressing spiritual and emotional needs of vulnerable women. Laura has also written for the international magazine El Centinela.

In 2024, Laura launched Health, Peace, and Trust: Thrive Amidst Adversity, a work uniting her extensive health knowledge with personal experience, emphasizing the importance of a spiritual connection with God for a full life. She hopes this book inspires readers to deepen their relationship with Christ, the Redeemer who transforms lives.

Laura is a distinguished member of the Guipil Academy: Write and Publish Your Passion.

For more information and contact:
laurie.fer@gmail.com

Bibliography

Books

Apelian, Nicole, Ph.D. (2023). The Holistic Guide to Wellness: Herbal Protocols for Common Ailments. Global Brother SRL.

Bulsiewicz, Will, M.D., MSCI. (2020). Fiber Fueled: The Plant-Based Gut Health Program for Losing Weight, Restoring Your Health, and Optimizing Your Microbiome. Avery.

Greef, Shanon & Willow, Natalie. (2024). Ancient Remedies Revived: A Complete Guide to Herbal Medicine and Natural Healing. GreenHaven Press.

McMillen, S.I., M.D., & Stern, David E., M.D. (2000). None of These Diseases: The Bible's Health Secrets for the 21st Century. Fleming H. Revell, a division of Baker Publishing Group, Grand Rapids, MI.

Otto, Sarah. (2023). The Inflammation Solution Summit: Restore Your Gut and Reclaim Your Life. Goodness Lover Pty Ltd.

Rojas Estapé, Marian. (2021). Encuentra tu persona vitamina: En la familia, en la pareja, en los amigos, en el trabajo. Editorial Espasa, Spain.

White, Ellen G. (1905). The Ministry of Healing. Pacific Press Publishing Association, CA.

Additional References

U.S. Centers for Medicare & Medicaid Services (CMS). Healthcare Spending Statistics. Retrieved from cms.gov.

Statista. Cosmetic Industry in the U.S. – Statistics & Facts. Retrieved from statista.com.

Netflix. (2024). You Are What You Eat: A Twin Experiment [Series].

Anderson, Kip & Kahn, Keegan. (2017). What the Health [Documentary]. Netflix.

Netflix. (2023). Live to 100: Secrets of the Blue Zones [Series].

Nayar, Anjali. (2024). Hack Your Health: The Secrets of Your Gut [Film]. Netflix.

Wikipedia. (2024, October 11). History of Medicine: A Study of Medical Practices Over Time. Retrieved from es.m.wikipedia.org.

Mayans, Carme. (2023). They Recreate the Perfume Cleopatra Used. Historia National Geographic. Retrieved from https://historia.nationalgeographic.com.es/a/reproducen-perfume-que-uso-cleopatra_14686

Biblical References

All Bible quotations were taken from the Reina-Valera 1960 version.

NOTES

NOTES

Made in the USA
Monee, IL
04 November 2026

67386688R00069